The Teaching of Ethics VII

Ethics and Engineering Curricula

Robert J. Baum

INSTITUTE OF
SOCIETY, ETHICS AND
THE LIFE
SCIENCES THE
HASTINGS
CENTER

Copyright © 1980 by the Institute of Society, Ethics and the Life Sciences

All rights reserved. No part of this book may be reproduced or transmitted in any form or by any means, electronic or mechanical, including photocopying, recording or by any information storage and retrieval system, without permission in writing from the Publisher.

The Hastings Center
Institute of Society, Ethics and the Life Sciences
360 Broadway
Hastings-on-Hudson, New York 10706

Library of Congress Cataloging in Publication Data

Baum, Robert J.
    Ethics and engineering curricula.
    (The Teaching of ethics ; 7)
    Bibliography: p.
    1. Engineering ethics—Study and teaching (Higher) I. Title. II. Series: Teaching of ethics; 7.
TA157.B343   174′.962   80-10099
ISBN 0-916558-12-6

Printed in the United States of America

Contents

Introduction		ix
I.	Definition of Engineering Ethics	1
II.	The Nature of the Engineering Profession	5
	A. Engineering Societies	6
	B. Codes of Ethics	7
III.	Characteristics of Engineering Students and Engineers	11
IV.	Engineering Education and "Real World" Experience	15
V.	Engineering Curricula	17
VI.	A Brief History of the Teaching of Engineering Ethics	21
VII.	Goals for the Teaching of Engineering Ethics	25
	A. Approaches to Setting Goals	28
	B. Basic Goals	31
VIII.	Who Should Teach Engineering Ethics?	33
	A. Appropriate Qualifications	33
	B. Team-Teaching	35
	C. The "Right" Instructor	37
IX.	Preparation for Teaching Engineering Ethics	39
X.	The Relation of Engineering Ethics to Other Fields	43
XI.	Topics for Consideration in Engineering Ethics Courses	47

XII.	Methods for Facilitating the Learning of Engineering Ethics	51
XIII.	Assessment of the Available Literature	55
	A. Textual Material	55
	B. Case Studies	57
XIV.	Research Needs and Priorities	61
	A. Whistle-blowing	61
	B. Informed Consent	67
	C. Discrimination and Affirmative Action	68
	D. Advertising	70
XV.	Conclusion	73
Notes		75
Bibliography		77

FOREWORD

A concern for the ethical instruction and formation of students has always been a part of American higher education. Yet that concern has by no means been uniform or free of controversy. The centrality of moral philosophy in the undergraduate curriculum during the mid-nineteenth century gave way later during that century to the first signs of increasing specialization of the disciplines. By the middle of the twentieth century, instruction in ethics had, by and large, become confined almost exclusively to departments of philosophy and religion. Efforts to introduce ethics teaching in the professional schools and elsewhere in the university often met with indifference or outright hostility.

The past decade has seen a remarkable resurgence of interest in the teaching of ethics at both the undergraduate and professional school levels. Beginning in 1977, The Hastings Center, with the support of the Rockefeller Brothers Fund and the Carnegie Corporation of New York, undertook a systematic study of the teaching of ethics in American higher education. Our concern focused on the extent and quality of that teaching, and on the main possibilities and problems posed by widespread efforts to find a more central and significant role for ethics in the curriculum.

As part of that project, a number of papers, studies, and monographs were commissioned. Moreover, in an attempt to gain some degree of consensus, the authors of those studies worked together as a group for a period of two years. The study presented here represents one outcome of the project. We hope and believe it will be helpful for those concerned to advance and deepen the teaching of ethics in higher education.

Daniel Callahan   Sissela Bok
Project Co-Directors
The Hastings Center
Project on the Teaching of Ethics

About the Author

Robert J. Baum

Robert J. Baum is Professor of Philosophy and Director of the Center for the Study of the Human Dimensions of Science and Technology at Rensselaer Polytechnic Institute. Dr. Baum is a member of the AAAS Committee on Scientific Freedom and Responsibility and director of the NEH National Project on Philosophy and Engineering Ethics. From 1974 to 1976 he was director of the National Science Foundation's Program on Ethical and Value Issues in Science and Technology. His six books and numerous papers are primarily on the topics of ethics, philosophy of science, logic, and technology and values.

Introduction

Engineering differs significantly from almost all other "service" professions in that its principal function is the design, production, and operation of specific physical objects, such as computers, automobiles, nuclear power plants, and pollution control systems—objects whose impact on individuals and on society as a whole is observable and assessable with little or no consideration of the role of the individual engineers who contributed to the creation of these objects. It is this feature of the profession, more than any other, which explains the fact that almost all courses presently being offered around the country in the general area of the ethical and social dimensions of engineering focus exclusively on the impacts of the *products* of the work of engineers.

Many ethical problems associated with the engineering profession have never been explicitly recognized or discussed to any significant degree outside the profession. The exceptions have generally been those cases that have been widely publicized and discussed in the media as "scandals" or in similarly dramatic ways. Some recent examples include the Goodrich "aircraft brake scandal" and the Bay Area Rapid Transit (BART) engineers' case. These cases are certainly significant, but their significance becomes magnified when it is recognized that they represent only the tip of the iceberg. Just as the Karen Ann Quinlan case called (and is still calling) public attention to a set of moral problems confronted by many families and physicians, the BART engineers' and other highly publicized cases involve a variety of issues faced by many other engineers in other contexts that affect directly or indirectly the health and welfare of large segments of the general public. Certainly the engineer who is designing, constructing, or testing a nuclear power plant, skyscraper, or passenger aircraft carries as heavy a burden of moral responsibility as the physician confronted with an irreversibly comatose patient sustained by artificial life-support equipment (equipment which, incidentally, was probably developed in part at least by engineers).

Not only are the moral issues confronting engineers as significant as those facing physicians, but they are also in some ways more difficult and complex, owing in part to the fact that most engineers are employees of (and entirely dependent for income on) a single firm or corporation, in contrast to most physicians' self-employed status. Many important and highly complex ethical issues associated with the engineering profession—such as problems of responsibilities to corporate and government employers, rights to freedom of speech (particularly in "whistle-blowing" cases), stealing trade secrets, ownership of patent rights, and certain kinds of conflict of interest—are encountered rarely in the medical profession.

I. A Definition of Engineering Ethics

The field of engineering ethics must be distinguished clearly from the related but quite different field of the study of the ethical (and other) impacts of technology. The latter field is concerned primarily with the study of the objects and organized systems of objects which fall under the general description, "technological." In contrast the field of engineering ethics is concerned with the actions and decisions made by persons, individually or collectively, who belong to the profession of engineering. While there is a close connection between these two fields, this relation between them is asymmetric in several ways. Since engineers are essential contributors in many ways to the creation of the technological entities that dominate our society, in considering their ethical responsibilities it is necessary also to take into account the impacts of the technologies which they have created. In contrast, it is often not essential or even particularly useful, to consider the role of engineers in the creation of a technology when assessing the impacts of that technology on society. (One exception might be the fact that the societal resources expended in training and supporting the engineers to develop a certain technology could be channeled in other directions, producing other impacts. But even in this case we aren't really talking about the responsibilities of *engineers,* but rather the responsibilities of society to support individuals and/or to train individuals who become engineers.)

A 1977 NSF–supported survey conducted by the American Association for the Advancement of Science[1] suggests that most colleges and universities in the United States today have at least one course (and many have more) that deal in some way with the ethical and value impacts of science and technology. These courses range from the very general (e.g., Technology Assessment, Technology and Values, Technology and Public Policy) to the relatively specific (e.g., The Impact of the Automobile on American Culture, Computers and Privacy, Mechanical Technology in Contemporary Society). The readings used in such courses, and the approaches most often taken by instructors, permit little, if any, serious consideration of the role of individual engineers or of the engineering profession as a whole in the control of the technologies whose impact is being studied. In other words, most of these courses focus on the questions of what the impacts of various technologies are and whether these impacts are good or bad; they don't look at the question of what engineers can or should do or what engineers' responsibilities are, especially with regard to the more negative impacts.

Although many of these courses are taught from the perspectives of a single discipline, a large number of them are interdisciplinary, involving persons from such departments as engineering, law, the social and behavioral sciences, and the humanities. Today, literally hundreds (possibly thousands) of these courses are offered within formally established programs, ranging from subdisciplinary programs on the philosophy of technology and the history of technology to broadly interdisciplinary technology studies and technology and public policy programs. Most of these courses are open to all students with no prerequisites. Some courses can be used to satisfy humanities or social science requirements for engineering students; others satisfy science requirements for humanities majors, and some can be used to satisfy either requirement.

Although such courses deal with many topics which are of more or less direct relevance to the ethical decisionmaking process of individual engineers (especially when any consequentialist theory of ethics is being applied), their quantity and diversity will not permit them to be explicitly treated in the remainder of this study. The main focus here will be restricted to the teaching of engineering ethics, where "engineering ethics" is defined quite narrowly as dealing with *judgments and decisions concerning the actions of engineers (individually or collectively) which involve moral principles of one sort or another.* (This definition is intended to include disputed cases where some persons believe a moral principle is involved, but others might hold that there is no moral tenet relevant to the case.) Normative judgments of the impacts of a technology or technologies will be discussed only as they relate to specific cases of ethical judgments concerning actions of engineers.

II. The Nature of the Engineering Profession

In contrast to law, medicine, and a number of other modern professions, the profession of engineering is not very easily defined, especially with regard to identifying its membership. No single set of necessary and sufficient conditions can be specified for gaining admission to the ranks of practicing engineers. It is not only unclear as to when a person enters the profession of engineering; it is even more vague as to when a person *leaves* the profession. Because of this definitional difficulty, estimates of the number of practicing engineers in the United States today range from less than 500,000 to more than 1.5 million. The most widely accepted figures are somewhere near the middle of this range. The problems of defining the profession are of relevance to a number of the central ethical issues, and also to the teaching of engineering ethics, so it will be worthwhile to look into this topic briefly.

While licensing and registration for engineers exists in all fifty states in the United States, a license is not required for engineering practice to the extent to which it is required for legal or medical practice. In fact, barely half of the minimal estimate of engineers in this country is registered by any of the states, and there is considerable debate even among members of the profession as to whether or not registration requirements should be broadened or tightened. At present it is possible to be employed as an engineer, and even to become licensed, without a degree in engineering. A number of persons employed as engineers have degrees in physics, chemistry, and other nonengineering disciplines, and some engineers have entered the profession through on-the-job training programs rather than through any formal degree program. In industry especially, an engineer is often any person whom management chooses to call an engineer.

It is at least as unclear as to when a person ceases to be an engineer. Statistical analyses generally quote a figure of between 50 percent and 75 percent as the percentage of individuals who started their careers in engineering positions and ultimately moved into management positions of one sort or another. (Of course, "management" is at least as vague a term as "engineer," so this judgment is doubly questionable.) Many of the "managers" supervise only small teams of other engineers on purely technical projects, and thus are doing what most would still consider to be basically engineering work. But the presidents of some of the largest corporations started their careers as engineers, and they still consider themselves to be engineers. In fact, many of these top executives serve as high-ranking officers in the various professional societies. Other individuals move from what are clearly engineering positions to positions in marketing, personnel, and other areas, again continuing to identify themselves as engineers.

Taking into account the basic problem of adequately defining the field of engineering in general, we shall, for the purposes of this discussion, restrict ourselves to a limited definition of the engineering profession as including persons who have completed at least a B.S. program in engineering (or engineering technology), and who are initially employed in positions that are described by the employers as engineering positions (which would include technical sales, quality control, and other non-R&D positions). This is not an entirely arbitrary delimitation of the field, since our main focus is to be on the teaching of ethics in college-level engineering curricula.

A. Engineering Societies

One feature of the engineering profession which is of some relevance to the teaching of engineering ethics is the structure and role of the professional engineering societies. Unlike medicine, law, and many other professions, there is no single "umbrella" organization that plays a dominant and central role for even a majority of the profession. The engineering societies are, for the most part, organized around the various engineering specialties, such as electrical, mechanical, civil, and chemical engineering. In addition, there are several general organizations, such as the National Society of Professional Engineers (NSPE) which are open to all holders of an engineering license. (None of these general organizations have as large a membership as the larger specialty-oriented societies.) Many of the societies are purely technical organizations, restricted to the holding of conferences and the publishing of journals and technical papers; as such, they often do not have codes of ethical conduct, nor do they consider discussions of ethical issues as appropriate for any of their meetings or publications. Most of the larger engineering societies are concerned with the advancement of the engineering profession in a broader sense, and they have formulated and set up minimal mechanisms for enforcing codes of ethics.

There have been continuing efforts during all of the more than one hundred-year history of the professional engineering societies to establish a single organization that would represent and serve the interests of all engineers, but none of these efforts has been successful to date. The most significant unifying organization at present is the Engineers' Council for Professional Development (ECPD), which is a coalition of sixteen of the major specialty societies. The ECPD is primarily an administrative organization responsible for accrediting professional engineering degree programs. It has, however, also taken a leadership position with regard to formulating a generally acceptable code of ethics during the last three decades. It is significant that its model code has been adopted (sometimes with modifications) by only about 30 of the 150 or more engineering societies. It is also significant that the Institute of Electrical and Electronics Engineers (IEEE), which has the largest membership of all of the engineering societies, has a code of ethics which differs most radically from the ECPD code, even though IEEE is one of the members of ECPD.

B. Codes of Ethics

A definitive history of the early development of the codes of ethics of the various engineering societies is provided in Edwin

Layton, Jr.'s *Revolt of the Engineers,* which is "must" reading for anyone interested in teaching or writing on any aspect of engineering ethics. However, this study only goes up to the period of the Second World War, and thus does not provide the complete perspective essential for understanding the current codes of the engineering societies. There are two features of the present codes that are of considerable significance in the teaching of engineering ethics. The first is the fact that the codes are almost totally restricted to considerations of the responsibilities of engineers, i.e., there is little attention paid in any of the codes to the *rights* of engineers or to the protection of engineers who adhere to the specified responsibilities. The second feature of the codes they have changed significantly in certain ways over the years. It is important for all those involved in the teaching of engineering ethics to recognize this development of the codes of the engineering societies, for it is not adequately noted in any of the existing literature. The codes have, in fact, gone through a quite significant change over the sixty or so years since their first formulations.

The first code adopted by an American engineering society was that of the American Institute of Electrical Engineers, which was adopted in 1912. This code, which served as the model for the codes of most of the other engineering societies, had the following as its principal stipulation: "The engineer should consider the protection of a client's or employer's interests his first professional obligation, and therefore should avoid every act contrary to this duty." There is no mention in any of these early codes of the engineer's responsibility to the general public beyond statements such as the following: "The engineer should endeavor to assist the public to a fair and correct general understanding of engineering matters, to extend the general knowledge of engineering, and to discourage the appearance of untrue, unfair, or exaggerated statements on engineering subjects in the press or elsewhere. . . . "

It was not until 1947 that a major reformulation of the codes of engineering ethics was made by the various societies, using the new Canons of Ethics for Engineers of the Engineers' Council for Professional Development as the model. This code begins to show a concern for the public welfare, indicating that this should be weighed against the responsibilities to employers and clients. The formulation in the code is as follows: "As the keystone of professional conduct is integrity, the engineer will discharge his duties with fidelity to the public, his employers, and clients, and with fairness and impartiality to all. It is his duty to interest himself in public welfare and to be ready to apply his special knowledge for the benefit of mankind." The duty of the engineer to the employer or client is no longer "paramount," and it is now specified only that "the engineer will act in professional matters for each client or employer as a faithful agent or trustee." It is also explicitly stated in this 1947 code that the engineer "will have due regard for the safety of life and health of the public and employees who may be affected by the work for which he is responsible." It was not until 1974 that the ECPD canons were revised to where they now read, "Engineers shall hold paramount the safety, health and welfare of the public in the performance of their professional duties." The ECPD canons have been adopted by many of the major engineering societies (including ASCE, ASME and AIChE) with at most minor modifications. Only the Institute of Electrical and Electronics Engineers has developed a significantly different code. That code does not specify that the public interest must be held *paramount,* but it does indicate that an engineer's responsibility to employers or clients are *limited* by their responsibilities to "protect the safety, health and welfare of the public. . . . "

Not only have many engineers failed to adequately recognize these changes in their codes over the years, but with the exception of a few individuals they have not seriously discussed these basic principles in any critical analytical way.[2] The most articulate critic of the latest formulation of the codes is Samuel Florman, who has argued in his book, *The Existential Pleasures of Engineering*,[3] and a number of other articles[4] that since engineers are no more qualified than any other individual citizens to determine what is, in fact, in the best interest of the general public that they therefore have no special responsibilities whatsoever to protect the general welfare (whatever it may be). At times, Florman seems to qualify this apparently extreme position to read that the engineer has a responsibility to do whatever the public says is in its best interest, but his arguments are generally quite vague and general, and his position is difficult to pin down.

Although Florman and several others have provided a significant stimulus for the initiation of open debate and discussion of issues such as this, there really is almost no literature dealing with the topic in any way which is usable in the classroom. The lack of such a literature is one bit of evidence that few engineering faculty are sufficiently involved in the *critical* analysis of the fundamental tenets of their codes to be able to deal with these issues in a sufficiently sophisticated way in an engineering ethics course. It is in this context that consideration of the codes of ethics of other professions can provide useful insights, and that the general modes of ethical analysis from philosophy can be applied in helpful ways.

It should also be noted briefly that although none of the codes presently discusses the *rights* (as distinct from responsibilities) of engineers, most of the major engineering societies have endorsed the "Guidelines to Professional Employment for Engineers and Scientists" which do spell out a number of (mostly economic) rights of engineers employed in large organizations. However, these guidelines still do not address the issue of the protection of engineers from reprisal for taking actions in conformity with the codes of ethics. This problem has been addressed head-on to date only by the IEEE, which has incorporated a special set of procedures for censuring employers in its constitution. The first case under these procedures has just been processed, with the issuing of a public statement criticizing an employer's actions in firing a member of IEEE who was acting in compliance with the IEEE code.[5]

III. Characteristics of Engineering Students and Engineers

Engineers as a group, including students as well as professional practitioners, are a distinctively homogeneous group, even in comparison with more restricted professions, such as law and medicine. An understanding and appreciation of some of the basic characteristics shared by the bulk of the membership of the profession (and engineering student body) is necessary for anyone interested in discussing ethical issues with members of the group. The profession is dominantly male, with fewer than 10 percent of the graduating engineers in 1979 being female and a much smaller percentage of the practitioners being female. (More than 20 percent of the students entering in 1979 were women.) The self-image of students, even the latest class of undergraduates, is such that many male students still perceive engineering as being a male profession. The majority of engineers and engineering students come from lower middle-class and upper lower-class economic backgrounds, and they regard engineering as a profession that will provide them with upward mobility on the socioeconomic ladder. Engineers generally have a preference for working with things rather than with people, and they are uncomfortable dealing with and expressing personal feelings and emotions. Engineering is the only field where students in all subareas (electrical, chemical, and so on) have average scores of less than 500 on the verbal part of the Graduate Record Examinations. Engineers also have the highest average quantitative scores on most standardized tests.

These basic traits are manifested in a variety of ways in the context of engineering ethics courses. Almost all of the students come into an engineering ethics course sharing a fundamental assumption about the nature of ethics which is, in turn, grounded in a deeper set of assumptions concerning the fact/value distinction. Engineering students are very comfortable talking about reified concepts of facts and values, tacitly accepting the corresponding ontological model that places the "facts" in the "external" world and the "values" in the individual person. This correlates with their epistemological assumption that there is a clear and sharp distinction between factual knowledge and value-laden knowledge, with factual knowledge being "objective" and value-laden knowledge being "subjective." Most engineering students assume that ethical knowledge is value-laden and totally "subjective," connected only to personal intuitions belonging to individual persons. Most of them also assume that science is concerned with factual knowledge about the external world and thus is totally objective. Since ethical knowledge is subjective and based on internal intuitions of individual persons, they believe that there is no basis whatsoever for resolving disagreements among individuals concerning ethical matters.

A thoroughgoing ethical relativism is therefore a common starting point for many engineering students at the beginning of an ethics course. But paradoxically, many engineering students are ethical absolutists, although they also share the relativists' assumption that there is no rational basis for comparing ethical positions, and they are unwilling (and/or unable) to present any arguments in defense of their own principles. They simply assert that they are true absolutely and apply to all individuals (not just themselves). At best, their position can be characterized as one of "naive intuitionism."

It is thus imperative that the instructor work through the whole set of epistemological and ontological presuppositions—doing almost a minicourse on philosophy of science as well as metaethics—in order even to begin to communicate with students concerning the possibility of doing ethics in any sense that involves the resolution of ethical conflicts or disagreements. The students are not necessarily easy to convince of the difficulties with their naive positivist position, but at least they must be made aware that the instructor is bringing a different set of epistemological and ontological presuppositions and assumptions into the course.

Associated with their stronger skills and abilities in quantitative rather than verbal areas, and also with their basic preferences for working with things rather than with people, engineering students often have a great deal of difficulty analyzing ethical cases as problems involving human interactions, especially with regard to identifying some of the more subtle aspects of the cases and considering some of the more humanistic alternative courses of action that might be open to an individual in a given situation. They are also more inclined to look for "technological fixes" to ethical problems and dilemmas, showing little patience with psychosocial criticisms of their solutions. Engineering students also have a tendency to view people in simple black and white terms as good persons or bad persons, and their analyses of cases usually begin with simplistic categorizations of the individuals involved into these two groups. Thus a course dealing with ethical issues for engineering students often has to involve consideration of psychological and sociopolitical factors to a much greater degree than might an ethics course oriented toward humanities or social science students.

Unfortunately, engineers (both academic and nonacademic) and engineering students for the most part have little awareness of the history of engineering or of the social and political structure of the profession, especially as these relate to the various ethical issues associated with the profession. This fact is of relevance to the question of qualifications for teaching engineering ethics and the contents of engineering ethics courses, and will be discussed further in appropriate contexts later.

IV. Engineering Education and "Real World" Experience

In sharp contrast to lawyers and physicians, most engineers (probably well over 75 percent) are employed in relatively large industrial or governmental organizations. Similarly, most engineering students are planning to enter employment in such organizations upon graduation. It is significant, and oft-noted by corporate managers, that most engineering degree programs are modeled on science programs, with a focus on problem-solving skills and basic theoretical knowledge. Little effort is made to simulate "real" employment situations, even in the technical courses. There is nothing in most engineering programs comparable to the clinical experience provided in most medical and an increasing number of law programs. Even the senior design courses, which are intended to provide at least a simulation of "real" job assignments, generally do not consider the human components of a real job situation where the engineer has to deal with managers or clients, under specific contractual terms. The only opportunity for an engineering student to get first-hand experience with nonacademic engineering working environments is in a summer job or a co-op program. Students who have had such experience perform quite differently in engineering ethics courses from students who have not had such experience.

Insofar as familiarity with industrial and governmental employment situations is of considerable relevance to the understanding of the fundamental ethical issues of engineering, and since in most universities relatively few students participate in co-op programs, it is necessary to try to compensate for this lack of experience in engineering ethics courses. The use of case studies *might* be helpful in dealing with students lacking in relevant first-hand experience, but there are many pitfalls in using them in such situations, some of the more significant of which will be dealt with in a later section. One problem with all of the currently available case studies is that they present only situations in which major difficulties have been identified, and presumably they are not very representative of the "normal" situations in which most engineers are employed. The lack of first-hand experience on the part of the students makes it necessary to provide a number of "positive" cases in which potential difficulties have been worked out satisfactorily within the organizational framework, in order to give students an adequate overall perspective on the exceptional cases in which engineers had to resort to extreme strategies such as whistle-blowing or resignation.

V. Engineering Curricula

Although the first two engineering schools in the United States were established early in the nineteenth century (West Point Military Academy and Rensselaer Polytechnic Institute), formal engineering curricula did not attract significant numbers of students until after 1870. Today, over 300,000 students are enrolled in B.S. programs (including Engineering Technology) in over 250 colleges and universities. Another 40,000 students are enrolled in graduate degree programs in engineering. Few of these students are being exposed to any rigorous or systematic treatment of the ethical issues which they will have to address as part of their professional careers, but the situation may be in the process of significant and rapid change. Few engineering faculty, and at most a handful of nonengineers, were devoting any serious attention to this field prior to 1976. As of mid-1979, more than one hundred persons, almost half of whom are nonengineers, have become engaged in a variety of teaching and research activities in the area of engineering ethics.

In engineering, as with most disciplines and professions, the structures of the various curricula are controlled almost exclusively by the academic members of the profession, with the main outside forces coming from the larger academic community and from the industries who employ graduates of the programs. At present, the basic requirements for accredited B.S. programs include a minimum of six courses in the humanities and social sciences, and a substantial number of courses in mathematics, physics, and chemistry, in addition to general and special engineering courses. For the most part, engineering faculty play an active role in advising students, and they usually advise them to take as many technical courses as possible and not to waste their time on "soft" courses. Students usually follow this advice.

There have been a number of ongoing debates concerning the overall structure of engineering education that are relevant in various ways to the inclusion of ethics courses in the curriculum. Arguments have been offered over the years on behalf of "professionalizing" engineering education along the models of medicine and law. These proposals argue that there should be a preprofessional program of two, three, or four years, which would be basically a liberal education undergraduate program with certain required science and general engineering courses, to be followed by a two-, three-, or four-year professional specialization program similar to the law school and medical school curricula. Such a lengthening of the curriculum from four to a minimum of five to seven years would make room for both general ethics courses at the preprofessional level and a specialized engineering ethics course at the professional level. At present, however, there are few indications that such programs will be established in the near future.

The present trend is in the direction of an extension of the program from four to five years, but only by the route of providing an automatic fifth year masters' program. This alternative has no effect on the undergraduate program and provides no additional opportunities for consideration of nontechnical topics, such as ethics, during the fifth year (that is, the M.S. graduate degree year). In brief, the argument for extending the duration of the basic engineering program is tied primarily to the claim of the engineering faculty that there is insufficient time in the existing four-year program to cover all of the *technical* topics that are essential before a person can enter professional practice as an engineer. Little real concern has been expressed by most engineering faculty that the six to eight humanities and social science courses may possibly be less than optimal for persons entering the profession. It is for this reason that most proposals for requiring an engineering ethics course suggest that this course be counted as satisfying one of the required six humanities and social science electives. This assumption that the existing curriculum is already too short to fit in all the necessary technical material also explains in part why few engineering faculty will incorporate any significant discussion of ethical issues into their regular engineering courses.

Since there is some reason to believe that "outsiders" will become more actively involved in the setting of curricula and other requirements for admission to the various professions in the future, it is worth speculating as to what effect this may have on engineering curricula. There is already considerable disagreement between the academic engineers and the nonacademic engineers as to exactly what is essential for a basic engineering education. Even the increased involvement of nonacademic professional engineers in the design of engineering curricula could result in radical changes in those curricula. It is quite likely that a curriculum designed by individuals trained as engineers who are now corporate managers would involve the deletion of numerous courses that are now required, and replacement of those technical courses, not only by new kinds of technical courses, but even more so by courses on topics such as technical writing, communication skills, managerial economics, and industrial psychology.

The important issue raised by the present conflict between academic engineers and nonacademic engineers is that there is at least some reason to believe that the curricula are not as tight and inflexible as academic engineers want their colleagues—especially in the humanities and social sciences—to believe. In any event, the increased involvement of nonacademic engineers and the new involvement of nonengineers in the design of engineering curricula would at least result in a significant reevaluation of the potential amount of space in existing undergraduate curricula that might be made available for nonengineering courses, including courses on ethics. Even if the discussion of the necessary quantity of technical courses supports the status quo, the nonengineers might very well press for an expansion of the professional program into additional years so that additional nontechnical courses, including ethics and related relevant topics, can be covered. It cannot be emphasized too strongly that, as the curricula stand now, given the basic experience and orientation of typical undergraduate students in engineering, a great deal of time in any ethics course must be spent in dealing with peripheral issues that could or should be dealt with in more depth in other courses, such as psychology, political science, history, sociology, and communications.

VI. A Brief History of the Teaching of Engineering Ethics

A survey of the engineering journals and other materials suggests that there has been a small and relatively constant percentage of engineering faculty who have been trying to incorporate consideration of ethical issues into engineering curricula in various ways over the past hundred years. Articles have appeared regularly over this period in various engineering publications decrying the morally questionable practices of some members of the profession and calling for various steps to be taken in the curriculum to try to encourage more ethical behavior on the part of future engineers. For the most part, discussions of ethics have been tied almost inseparably to the topic of professionalism in engineering, and few courses have ever been offered that dealt solely with ethics. Even courses described as being ethics courses usually dealt primarily with questions of legality and professional protocol. For example, a 1955 article in the *Journal of Engineering Education* listed the "four basic ethical problems [which] face the profession" as giving kickbacks to get contracts, accepting commissions from contractors or manufacturers, price bidding, and undercutting the professional reputation of another engineer. The people teaching these courses were generally senior engineering faculty.

Two studies conducted over the past twenty years by the National Society of Professional Engineers provide some perspective on the place of the teaching of ethics in the traditional engineering curriculum. The results of a survey published in 1963 by NSPE[6] indicated that approximately 25 percent of the engineering schools responding to a questionnaire used some kind of formal instruction to deal with certain issues in the general area of "professionalism" in engineering. The report indicated that this was most commonly done in some kind of a freshmen orientation course or program. Little was done in these courses on the topic of ethics beyond discussion (*not* critical analysis) of one or more of the professional society codes. A second survey, conducted in 1971, indicated that fewer faculty were dealing with the topics of ethics and professionalism in any formal way.[7] The engineering faculty respondents reported that they were instead trying to address these issues in the context of faculty-student interactions outside the classroom, and through a process of role-modeling, rather than by formal instruction. Once again, "ethics" seemed to mean to most of the faculty little more than familiarity with the code of ethics of one or more of the professional societies.

In 1975, the NSPE published a questionnaire in its monthly magazine, *Professional Engineer*, asking engineers to rate their undergraduate education in professionalism. Only 300 engineers (out of more than 60,000 recipients of the magazine) responded, and they indicated that they felt that their undergraduate training in the general area of professionalism was inadequate. In 1976, NSPE published a new report based on information about formal courses in professionalism solicited from the deans of all accredited engineering schools.[8] Out of almost 250 accredited schools, 128 responded, and only 13 reported that professionalism was taught as a separate course in their program. The course outlines that were submitted indicate that at most 10 to 15 percent of the time in any of these courses was devoted to the consideration of anything that might be considered to be ethical issues, and once again, this was limited almost entirely to noncritical readings of the codes of the various societies. Other topics covered in such courses included history of the profession, patent policies, economics, relations to peers, the structure of the engineering profession, community relations, and legal aspects of professional activity (contracts, liability, etc.).

In its 1976 report on the 1974 survey of the deans, NSPE made specific recommendations concerning the inclusion of courses on professionalism in engineering curricula. They recommended basically that all other engineering schools develop courses resembling in some way the kinds of courses identified at the thirteen schools reporting courses in the 1974 survey. It was tacitly recommended in the report that the course be run by engineers and that most, if not all, visiting lecturers be engineering faculty or nonacademic engineers.

With the exception of a few historians, the most notable being Edwin Layton, Jr., few, if any, philosophers, lawyers, social scientists, or other scholars doing work relevant to the topic of engineering ethics were ever involved in any significant way in any course on this topic offered in engineering departments prior to 1975. Moreover, there is no evidence that there were any courses specifically focusing on engineering ethics being offered outside engineering departments by nonengineering faculty prior to 1975, other than one course in the Philosophy Department at Rensselaer Polytechnic Institute. Beginning around 1976, in part as a result of the attention being drawn to this area by new funding programs at the National Science Foundation and the National Endowment for the Humanities, a number of humanists and social scientists began to become involved in a variety of activities dealing with engineering ethics, including the development of new courses, in collaboration both with engineering faculty and in the basic humanities and social sciences disciplines.[9] As of 1979, there were probably thirty to forty such courses being offered,[10] and at least fifty more courses are in various stages of planning and should be offered within one to two years.

The accrediting organization for engineering curricula, the Engineers' Council for Professional Development, is in the process of considering whether to formally require a course on ethics and professionalism in all accredited engineering curricula. This debate within the accrediting organization is itself providing a stimulus to engineering schools to develop such courses on a nonmandatory basis, and it is likely that many engineering schools will continue with such courses and/or add new courses even if they are not ultimately required for accreditation.

VII. Goals for the Teaching of Engineering Ethics

Although there is increasing agreement between engineers and nonengineers that "ethics" should be dealt with in some way in engineering curricula, there is little agreement as to what the goal of such activity should be or what content should be included. Acknowledging that different schools already have various approaches and probably will continue to do so, the report on the NSPE 1974 survey recommends that all engineering schools should have at least one course which has the following goal:

> [T]o provide the student with some orientation to the current status, practice and problems of the engineering profession and its relationship with the rest of society.[11]

Such a statement is of course so general as to be of little operational value in terms of knowing what is intended concerning ethical issues in particular. However, the general literature dealing with this topic in engineering publications is somewhat more explicit, and it does give an idea of what is probably the consensus opinion among engineers as to why ethics should be taught in the engineering curriculum and what the goals of such courses should be.

One such article, appended to the NSPE report, stipulates that two goals of engineering education are: (1) "to give every engineering graduate basic values of professional conduct"; and (2) "to give every engineering graduate a sense of responsibility for the exercise of care and objectivity in the application of the knowledge and technical expertise entrusted to him."[12] A reading of the many articles by engineers containing statements of this sort indicates that most of them interpret phrases such as "basic values of professional conduct" as referring to conformity with the basic tenets of the codes of the various engineering societies that deal with general principles concerning relations with employers and fellow professionals, as well as with the general public. Further, there is nowhere in the literature any indication that engineering education should even involve *critical* analysis or evaluation of such codes, with the exception of possibly considering the problems of applying some of the more vague or ambiguous sections of the codes to concrete situations.

Discussions by engineers of the teaching of ethics almost always express, tacitly if not explicitly, the ultimate purpose of making future engineers *behave* more ethically in their professional capacities, again where "more ethically" is understood to mean "in compliance with the tenets of the professional society codes."

In brief, engineers on the whole do not perceive the goal of ethics teaching as being significantly different from what physicians, lawyers, or educators in general see it to be. In essence, they are all believers in the "educational fix" approach to basic social problems. They are believers in the myth that it is possible to correct the difficult problems created by technology, economic injustice, and even fundamental human nature by running students through a one-semester course on this or that or something else. There is little recognition that students come into the curriculum with a deeply ingrained set of values and basic outlooks and beliefs about all aspects of the nature of reality, including human nature, social values, and the meaning of life in general. They are sometimes aware that students self-select into the engineering profession, and are a highly homogeneous group in terms of their basic values and personality characteristics, but most engineering educators do not connect this fact with the question of the basic ethical values of the members of the profession. Thus, it is not generally recognized that if the goal of changing the ethical orientation of engineers is a desirable goal in one way or another, the appropriate and probably only workable method for doing this would be to change the standard of admission into the profession to attract and encourage a more socially concerned type of individual, rather than to try to take individuals who do not have a great deal of social concern and instill such a concern in them at a relatively late age. Even if this latter could be done, and even if it were deemed to be socially desirable, the proposal to modify the values, attitudes, and/or behavior of these young adults would raise a serious set of questions concerning behavior modification, behavior control, and so on. The lack of feasibility of such a strategy fortunately makes such considerations unnecessary at the present time.

The issue of possible changes in standards and procedures for admission to the profession of engineering itself involves a number of significant substantive issues, none of which are discussed anywhere in the literature and none of which have been discussed in the context of engineering ethics or the teaching of ethics in the engineering curricula. At present, as with almost all other professions, the degree standards and requirements for engineering curricula, as well as the licensing and registration procedures in most states, are controlled by members of the profession itself. These policies and procedures are being challenged at both the federal and state level by both regulative and legislative bodies, but the ramifications of this practice have not been deeply explored from an ethical point of view.[13]

The basic argument in support of the status quo—the control over standards and procedures for admission to the profession *by* members of the profession—is that only members of the profession have the expertise necessary for making the necessary decisions in a sufficiently intelligent way to protect the interests of the general public. This paternalistic position assumes that it is impossible for members of the profession to explain adequately to laypersons the relevant considerations in the establishment of such policies so as to permit laypersons to participate in the decisionmaking process. (It is almost always also assumed that the incapacity is that of the layperson to understand, rather than the incapacity of the professional to explain the issues to the layperson.)

The basic argument against the status quo—and in support of the involvement of laypersons in the setting of standards and procedures for admission to the profession—is that there is an inherent inescapable conflict of interest on the part of members of the profession which can only be avoided by the participation of "outsiders" in the policymaking process. The control of admission to professional practice not only sets minimal standards of competence which are asserted to protect the public from incompetents and charlatans; the control of admission also protects members of the profession from an excessive supply of competitors for their services. Proponents of lay control of standard-setting policy assert that the only cause of any failure on the part of laypersons to understand the fundamental issues is the failure of the members of the profession to make an adequate effort to communicate this information to laypersons.

The resemblance between this issue and the question of the decisionmaking process in the areas of medical treatment and human experimentation are sufficiently great to suggest that the burden of proof should be placed on members of the profession to demonstrate that laypersons cannot participate intelligently in this process, and to assume that laypersons do have such a right unless the engineers can prove otherwise. But this is a matter which requires discussion and dialogue between the profession and the general public, rather than proof, demonstration, and/or adversary procedures. The goal should be one of mutual understanding and agreement, with a recognition of fundamental common interests being shared by both groups. The point is being made here primarily to bring out as sharply as possible two contrasting concepts of goals for the teaching of ethics in the engineering curriculum. One of these involves the setting of goals by the "experts"; the other involves the setting of the goals by the clientele being served or even by the society at large, in collaboration with appropriate "experts."

A. Approaches to Setting Goals

The first approach, that in which the goals of the ethics component in the engineering curriculum are set by "experts," is that which is almost universally used today. Insofar as the engineering profession itself has almost total control over the required components of the engineering curriculum, it possesses and exercises the *de facto* and *de jure* control over any required ethics components. There is no indication that there has ever been any serious consideration of involving "outsiders" in the deliberations concerning any aspect of the engineering curriculum, including the ethics component. The result of this control over the decision-making process is the situation outlined earlier, namely, the existence of relatively few ethics courses anywhere in engineering curricula and no courses with any significant philosophical sophistication.

The one alternative to the control of this aspect of the curriculum by the engineering profession itself is the development of independent elective courses on engineering ethics outside the engineering schools, usually in philosophy or humanities departments. While these courses have a distinctly different tone and orientation from the courses developed and offered by engineers in the engineering curriculum, they still for the most part share the basic mode of goal-setting by "experts." The only difference in this respect is that the "experts" are now philosophers rather than engineers, and the expertise is measured in terms of competence in dealing with ethical theory as opposed to competence concerning the nature of engineering practice.

Neither of these approaches to the setting of goals for ethics courses in the engineering curriculum is any more appropriate or justifiable than the existing approach to the setting of standards and procedures for admission to the profession. From an ethical point of view, the burden of proof still rests on both engineers and philosophers to demonstrate that their clientele (i.e., engineering students) and/or the general public are not competent to make or at least to participate in the making of these decisions. Since neither empirical evidence nor any compelling a priori arguments have ever been presented that would support such a claim, it therefore follows that the question of what the goals of ethics teaching in the engineering curriculum should be must be left to the students and/or the general public, with at most input from and/or collaboration with the "expert" engineers and philosophers.

There is a second argument, based on some of the fundamental assumptions of philosophers involved in the teaching of ethics, that leads to the same conclusion. It is widely agreed among philosophers and others who are actively involved in the teaching of ethics in the engineering curriculum and in other contexts that one of the primary goals of the teaching of ethics should be to develop the students' sense of moral autonomy and responsibility, and a second major goal should be to make the instructor unnecessary. Students in general, and engineering students in particular, have a relatively weak sense of their own moral autonomy. They have been conditioned to look to their instructors for the answers to any and all questions associated with any given course. The very structure of the educational system, even at the college level, rewards their acquiescence in playing the passive submissive role when confronted by paternalistic authority figures with claims of expertise in any specific area.

It is paradoxical, and in practice, self-defeating, for an instructor to try to impose in an authoritarian way a sense of moral autonomy on the part of her or his students. It is exactly this kind of predicament that Thoreau was able to avoid only by insisting that he did not want to have students or disciples, because in the very process of putting themselves in the role of students they would be relinquishing their own sense (and actuality) of being autonomous moral agents. The Socratic method itself is only workable when the "student" is taking initiatives directly and personally, and is, in fact, the "aggressor" in the dialectical process. Although Socrates himself in many of the dialogues was deviously manipulative, the method which he advocated was one of the "teacher" being in an essentially passive mode, acting as a mirror in which the actively inquiring person could see her or his own reflection. In brief, if one's goal is the instilling of a sense of moral responsibility in students, then the only morally justifiable approach and the only possible way of achieving this goal is to avoid trying to set goals or objectives for the student, including goals or objectives for their ethics course. If an instructor feels compelled, for whatever reason, to unilaterally stipulate or mandate certain goals for a course on engineering ethics (or any other course), then, at the very least, he or she has an obligation to make these goals explicitly known to the students, and when possible should encourage students who don't accept the goals to drop the course. It is difficult to think of any conditions under which an instructor would be justified in having a "hidden agenda" or unstated goals for the course.

If one accepts that it is morally unjustifiable to impose the instructor's goals on students or that the imposition of the instructor's goals is counterproductive in attaining some of the most fundamental goals, then most of the presently existing procedures being practiced for setting goals for ethics courses in engineering (and all other) curricula are unacceptable in one way or another. Other options must therefore be considered and evaluated. Ideally, at least from the ethical point of view, the setting of the goals for ethics courses in engineering curricula should involve not only the "experts," engineers and philosophers, but also the clientele—both students who are directly involved in the courses and representatives of the general public who are footing the bill and who are the persons ultimately affected by actions of the engineers in their professional capacities. Insofar as this involves a radical change from the existing situation, a great deal must be done before the ideal can be fully actualized. A number of steps can be taken at a variety of different levels simultaneously, but independently, which will contribute to a movement in this direction.

B. Basic Goals

First, philosophers and engineers, as "experts" approaching the subject matter from different perspectives, can and should begin an increased dialogue with colleagues and members of the other group, both on the individual level and on the general professional society level. Great benefit would also be derived, if both groups were to increase the dialogue between their professional organizations and the larger society.

It is of particular importance that both engineers and philosophers should approach their courses in engineering ethics with a determination to involve the students as much as possible in the process of the setting of the goals for specific courses. In cases where this has already been attempted, it has proven to be surprisingly successful. Certainly, the philosophers and engineers should not abdicate their legitimate responsibilities as faculty of providing suggestions and evalutions and recommendations based on their knowledge and experience in this area, and it is appropriate, if not essential, that they contribute a list of possible goals for such courses to students and/or representatives of the general public who are involved in the decisionmaking process. The following list of goals included in the recommendations of *The Teaching of Ethics in Higher Education: A Report by The Hastings Center* are goals which certainly should be presented for consideration by anyone concerned with the setting of such goals for specific courses or curricula, but they are not exhaustive of all of the possibilities.[14] These basic goals, which should be considered for adoption by any engineering ethics class, are:

1. Stimulating the Moral Imagination.
2. Recognizing Ethical Issues.
3. Developing Analytical Skills.
4. Eliciting a Sense of Moral Obligation and Personal Responsibility.
5. Tolerating—and Resisting—Disagreement and Ambiguity.

VIII. Who Should Teach Engineering Ethics?

The obvious answer to this question in most academic fields would be simply "The persons most knowledgeable about the subject matter." However, because of the lack of attention given to this particular field to date, there are at most a handful of persons having even a minimal familiarity with the various issues to be considered sufficiently qualified to teach courses on engineering ethics. The question as to who should teach is therefore much more open, and, in effect, is essentially the question: "Who should be trained to teach engineering ethics?" In brief, the answer to this question is: "Anyone from any appropriately related field, including engineering, religion, philosophy, history, sociology, and management." The two more interesting and more complicated questions are: "Who is the most (and least) *appropriate* specialist for teaching the various kinds of engineering ethics courses?" and "How should people be prepared for teaching engineering ethics?" The remainder of this section will be devoted to the first question; section X will deal with the second question.

A. Appropriate Qualifications

The persons most appropriately qualified to develop themselves to teach engineering ethics are those who fall under the recently evolved title of "ethicist"—whether their original disciplinary base was philosophy, religion (or theology), history, or one of the social or behavioral sciences. Individuals who have experience in more specialized areas, such as biomedical ethics, are perhaps slightly more prepared to move into this field than others who have more general backgrounds in a specific discipline, but because of the significant differences between many of the issues in medicine and engineering, experience in medical ethics could even be a disadvantage for some persons. (The dissimilarities between engineering ethics and medical ethics will be discussed in section X.)

Other individuals who also have good potential for contributing to courses on engineering ethics include sociologists, psychologists, lawyers, and historians. Sociologists with training or experience in the area of the sociology of the professions, the sociology of the corporation, and organizational structures are especially qualified for dealing with certain issues in courses on engineering ethics. Psychologists with training and experience in industrial psychology, interpersonal communications, and psychological studies of values and attitudes are also particularly qualified for handling certain issues in engineering ethics. Lawyers or legal scholars who are best prepared for contributing to courses on engineering ethics include those with specializations such as patent law, product liability, government regulations (e.g., Toxic Substances Control Act, Clean Waters Act, and Federal Aviation Agency regulations), and especially the few lawyers who are knowledgeable about recent developments in the application of various statutes (e.g., the Sherman Antitrust Act) to federal regulation of the activities of professional societies. Historians who have studied the development of the professions in general, and the development of the profession of engineering in particular, as distinguished from historians of technology, also have particularly appropriate backgrounds for moving into the area of engineering ethics.

However paradoxical it may at first appear, the least appropriate persons for teaching courses on engineering ethics are engineering faculty. There are several reasons for this. First, engineering students take a minimum of twenty courses from engineering faculty as part of their undergraduate programs. Any sincerely interested faculty person can (and should) incorporate discussions of ethical issues into these other courses. (Some of the problems of doing so are discussed elsewhere.) Second, it is essential in the teaching of ethics to provide "outside" perspectives on the issues under consideration, something which cannot be done adequately by persons deeply involved in the profession. Also, the *average* academic engineer has a quite narrow range of involvement and experience in the field, possessing little or no first-hand experience in nonacademic employment contexts, which is where more than 90 percent of their students will be working. (Engineering faculty who consult are employed *by* but are not *in* government or industry.) Third, engineering faculty who are perceived by their peers (and students) as being "real" engineers must be actively involved in research in their area of specialization—but ethics is not considered to be a legitimate specialization in engineering. Thus, academic engineers who devote sufficient time and energy to doing the research necessary to be competent to teach a whole course on ethics will generally have to cut back on, or even give up entirely, their engineering research. When they do this, they are then perceived by their peer community as being "ex-engineers," with no more, and often less, status than a philosopher or other nonengineer who might be qualified to teach engineering ethics.

B. Team-Teaching

Given that few academic engineers would ever be appropriately qualified for teaching courses on engineering ethics (for the reasons specified above), and given also that today few nonengineers are adequately qualified for teaching such a course alone, what would be the most appropriate way of designing and offering a course on this topic? It would appear that at present, in most academic institutions, the appropriate approach to staffing a course on engineering ethics would be to take a team approach, with the primary responsibility for the course being assigned to the individual who is judged to be the most knowledgeable about ethical theory and the application of ethical theory to individual case situations. This individual should be given every opportunity to do the background study and preparation outlined in section X, and should also be given the opportunity to involve as many people as possible in the first offering of the course on a full-time basis. In other words, especially during the first offering of the course, it would be best from a substantive point of view to have all the individuals from the various disciplines in attendance at every session of the course, in order to broaden their own knowledge and understanding of the issues and enable them to contribute to the others' understanding of the field. It would be of minimal value, both for students and faculty, to have six or eight different teachers coming in and out of the course on a session-by-session basis.

There is no single topic for discussion in an engineering ethics course which can be dealt with solely from one disciplinary perspective. Insights into any of the general issues—licensing, whistle-blowing, advertising of professional services, and so on—and into specific cases can be provided from each and every one of the disciplinary perspectives. Although it is theoretically possible to structure a course in such a way that during one period a historian might lecture on the history of the codes of the engineering societies, and a sociologist might discuss the same topic in a separate lecture on a different day, and a philosopher could analyze the codes on a third day, this mode of discussion and analysis is not particularly "natural" or helpful to students. The ideal situation would be to have a minimally structured discussion of the codes with a historian, sociologist, philosopher, and possibly an engineer or two, and people from other disciplines as well, contributing as appropriate in a more integrated way. After the course has been run in this way one or two times, the principal instructor should have a sufficient sense of the ways in which the individuals from the other disciplines can be fit in, and he or she can then design the course in such a way that faculty who do not have full co-instructor status can be brought in for guest lectures at appropriate points and can have their presentations adequately integrated into the rest of the course. Ideally, it would be best if the course could always be run at a sufficient level of enrollment to permit a minimum of at least three instructors to receive full credit for team-teaching the course. Thus, it would be better if the course were offered only once a year, or even once every two years, with a sufficient number of students enrolled to justify the assignment of two to four instructors to the course, than to offer the course every semester with a small enrollment that would only permit one instructor. If the total enrollment of the course exceeds thirty students, it should be structured in such a way that at least one session each week will allow discussion among a small number of students.

For a variety of reasons, including the lack of first-hand experience by the students and the lack of a literature on "normal" engineering practice, every effort should be made to involve nonacademic practicing engineers in courses on engineering ethics, even if academic engineers are also involved. This is more easily done logistically; alumni and officers of various professional societies are usually more than willing to come to one or two class sessions. Their mere presence is not sufficient, however, to provide a solid educational experience, and considerable groundwork must be laid to prepare both the visitors and the students in advance, so that the guest's presentation and the students' questions will fit into the overall course in a more or less natural and substantial way.

C. The "Right" Instructor

Bowing to the reality of academic politics and economics, it must be recognized that in many, if not most, institutions, courses on engineering ethics will ultimately be left to a single individual to teach, with at best the possibility of bringing in instructors from other disciplines for guest lectures here and there during the semester. A course offered under such circumstances, although necessarily less than ideal, can be of considerable value if taught properly by the right person. The "right" person for teaching an engineering ethics course, essentially as the sole instructor, would be an individual whose background and training are in ethical theory and the application of ethical theory to everyday situations, and who has personally prepared herself or himself for offering such a course in the ways indicated earlier as being essential for qualifying oneself to carry out such a task. If taught alone by a single instructor, a course on engineering ethics should (again ideally) have a maximum of twenty-five to thirty students, since the essence of such a course as a useful educational experience for the students involves the opportunity to interact in a dialectical way with the instructor. A person teaching such a course should rely to a significant degree on the audiovisual resources available and also on guest lectures, particularly from a variety of nonacademic engineers. It would generally be more of a distraction and disruption to try to "plug-in" formal lectures by individuals from other disciplines, other than possibly a one-hour lecture near the beginning of the semester on the history of the profession and the history of the development of the codes of ethics (by an historian) and a one-hour lecture on the sociology of the engineering profession (by a sociologist or social psychologist).

IX. Preparation for Teaching Engineering Ethics

Anyone interested in teaching engineering ethics, from whatever disciplinary background, should be adequately prepared in the following ways. He/she should be minimally familiar with the history of the development of the profession of engineering in the United States, particularly the development of the professional societies and their codes of ethics. At the very least, this would involve a careful study of the definitive book on this topic, Edwin Layton's *Revolt of the Engineers*. They should also have some familiarity with some of the sociological and political aspects of the profession, such as those provided by the papers in Perrucci and Gerstl's *Profession Without Community: Engineers in American Society*. Familiarity with similar perspectives on at least one other major discipline (preferably law or medicine) should also be acquired, if possible. Knowledge of the legal issues associated with engineering, including patent policies, licensing, product liability, governmental regulation, and similar issues would prove useful in considering many topics from an ethical perspective.

Much of the most important background reading for persons preparing to work in the area of engineering ethics is the more general literature on directly related topics and issues. In particular, some study should be done in the area of professionalism, in which several valuable works have recently appeared. Especially noteworthy are Randall Collins's *The Credential Society: An Historical Sociology of Education and Stratification* and Magali Larson's *The Rise of Professionalism: A Sociological Analysis*. In addition to providing a historical (and in Larson's book, a crosscultural) perspective on the broad range of professions, both books make significant advances beyond the literature, which had not seen any major developments in almost fifty years. Another important set of readings is in the area of the rights and status of employees of large organizations (since most engineers are so employed). Two important books on this topic are David Ewing's *Freedom Inside the Organization* and Deena Weinstein's *Bureaucratic Opposition: Challenging Abuses at the Workplace*.

It goes almost without saying that anyone interested in teaching engineering ethics should acquire a familiarity with the full range of substantive topics that comprise the field (these will be enumerated below). It may be less obvious, but it is no less important, that anyone teaching such a course should also have reasonable competence in ethical theory and the application of ethical theory to concrete issues (casuistry). Ideally, an engineer or other person teaching engineering ethics should have the equivalent of at least one course in ethical theory—even if he or she is team-teaching the course with an ethicist.

Any nonengineers teaching an engineering ethics course should try to get a good sense of the basic educational framework within which their course exists. They should not only familiarize themselves with the catalog descriptions of the engineering courses and curricula, but they should also try to sit in on some engineering courses to get a feel for the general classroom ambience as it reflects the values and belief structures of the students and engineering faculty. Attendance at student chapter meetings of the various engineering societies can provide useful insights for teachers of engineering ethics.

Given the lack of awareness on the part of undergraduate engineering students of the nature of the "real world" situations in which they will ultimately be operating as professionals, it is particularly important that individuals preparing to teach courses on engineering ethics take as many steps as possible to develop at least a second-hand familiarity with the variety of employment situations for engineers. This should minimally involve the reading of several years' back issues of a variety of different engineering magazines and publications, and, when possible, attendance at local or national engineering society meetings. In addition, an adequately prepared instructor should have spent a reasonable amount of time talking informally with a variety of nonacademic engineers employed in different situations, including at least a few employed as line engineers in large corporations, several engineers employed at middle-management levels in various kinds of organizations, and if possible an engineer or two employed by a governmental agency and/or a private consulting engineer. An ideal situation would be for the nonengineer to participate in a short course, workshop, or institute on engineering ethics, which includes among the participants at least a few nonacademic engineers. An alternative to talking directly with nonacademic engineers would be at least to spend some time talking with engineering faculty who were previously employed at some time in industry or government or private practice. However, it should be recognized that the perspective of the *former* nonacademic engineer is the perspective of someone who quite possibly did not find nonacademic employment to be satisfactory or satisfying, and thus it may be a quite different perspective from that of individuals who remain in the nonacademic organization and who find it a more or less satisfying experience.

X. The Relation of Engineering Ethics to Other Fields

There have been several indications in the preceding discussion that the field of engineering ethics shares certain similarities with, and is different in some ways from, a number of other newly developing areas of study. It is important to understand these similarities and differences as well as possible, in order to identify areas of potential support, additional resources and collaboration, and to develop a sensitivity towards potential pitfalls and problems.

The relation between technology assessment and engineering ethics was outlined in section I. There is a similar, and potentially more significant, relation between engineering ethics and the newly developing field of environmental ethics. The relation is similar, because like technology assessment, environmental ethics is primarily concerned with evaluating various impacts of specific technologies and of technology in general. The relation to engineering ethics is potentially closer for two reasons. First, the development of the field of environmental ethics has relied to a significant degree on the contributions of a number of competent and established philosophers and other humanists. (In contrast, the technology assessment field has been dominated by engineers and scientists since its inception more than a decade ago.) Also, environmental ethics is (by definition) focused on ethical issues, whereas the field of technology assessment usually deals with more general value judgments (and to a startling degree even tries to avoid dealing with value judgments at all!).

The field of environmental ethics is also at least as new as that of engineering ethics, and both are in a relatively uncertain position as regards their long-term viability. There is a potentially strong synergy that could be established between the two fields, but no real ties have yet been made. This is an area which persons in both fields should explore carefully.

An even newer and less developed field, but one which is of equally close proximity to engineering ethics, is that of computer ethics. Most engineers today use computers in their work, and many engineers are involved in the design and production of them. The primary work in this field has been led by computer specialist Donn Parker at Stanford Research Institute, and he has made considerable efforts to involve philosophers, social scientists, and others in his projects. He has also compiled a substantial (computerized) bibliography on this topic. The general issues in this field fall under the two categories of privacy and white-collar crime. Some courses on computer ethics are now being tested on an experimental basis.[15]

The field of engineering ethics has been developed along a track parallel with the field of business ethics. Philosophers (and other "outsiders") only began doing serious work in each field within the last five years, and new courses, texts, and other teaching tools have begun to proliferate nationally in both fields since 1977. Some of the films and texts on business ethics are very useful for courses in engineering ethics, and materials on engineering ethics have been used in business ethics courses. There are several significant potential connections between these fields, the most obvious arising from the fact (referred to earlier) that a majority of working engineers are employed by large corporations and that many of the top corporate executives began their careers as line-engineers. Also, many self-employed consulting engineers are incorporated and do much of their work under contract to corporations. But beyond these surface connections, there are many other important ways in which the fields of business and engineering ethics can and should be related. For example, many of the recent government actions to weaken the self-regulation of the various professions (discussed briefly in section VII) are grounded in the landmark 1974 Supreme Court ruling (*Goldfarb* v. *Fairfax County Bar*) that professional societies fall within the scope of the Sherman Antitrust Act. This has reopened a whole set of issues for serious consideration by engineers (as well as lawyers, physicians, and others) at the ethical as well as the legal level. These issues include advertising of services and competitive price bidding, both of which are important topics in business ethics. Philosophers, lawyers, and other scholars have much to contribute, not only to the increased understanding of these issues, but also to the development of implementable policies and standards. Further treatment of this topic will be given in section XIV.

In that an increasing number of engineers work directly for government agencies (from the local to the federal levels), and that many engineers in industry (and not just aerospace) are working on government contracts, there are some important ethical issues that might also be considered under the heading of public administration ethics and that relate to a lesser degree to the field of ethics and public policy. Since these potentially related fields are so little developed at present, the opportunities can only be pointed out in the most general way. It is to be hoped that they will soon be more fully explored.

The most obvious potential resource for the newly developing field of engineering ethics is the original area of applied ethics—medical ethics. Certainly much of what has been done to date toward the establishment of engineering ethics as a new field of interdisciplinary study has drawn on the experience of, and been modeled on, the successes of the field of medical ethics. There are two caveats that must be made concerning future relations between these fields.

The first warning is that there are some significant differences (several of which have been pointed out earlier) between the fields, where little may be transferable from one to the other. In some cases, the differences are subtle and even deceptive. While few (if any) engineers are self-employed in the manner of physicians, with a large number of individual clients, many (probably most) engineers are in situations more similar in various ways to those of nurses and members of the allied health-care fields. Thus, although engineers have traditionally striven for the status of and modeled their ethics codes on those of physicians, philosophers, and others working on engineering ethics should be open to looking for significant parallels with other elements of the medical field.

The second caveat is that because of its head start of more than ten years and the high quality of work done to date, there is a natural tendency to place medical ethics on a pedestal by those using it as a model for new fields, such as engineering ethics. In this position, it is perceived as a resource to be drawn from, but there is little sense that much can be contributed in return. It is important for the future health and development of both fields that the potential contributions of the new one to the established one be recognized and encouraged from the beginning. This need not be an artificial, contrived, or condescending act on the part of the scholars in medical ethics or a pretension to grandeur on the part of the engineering ethics scholars. There *are* numerous new issues to be raised and insights to be offered to medical ethics from engineering ethics, only a few of which can be listed here.

It has been pointed out several times earlier that the codes of ethics of the various engineering societies have been subjected to (at best) minimal critical analysis. This is also true of medical codes,[16] and since the engineering codes were originally modeled on medical codes, much of any original critical analysis that might be done on engineering codes will also provide insights into medical codes. Also, a number of basic issues have never been treated by medical ethics scholars—including advertising of services, setting of fee schedules, and control of access to the profession; thus this work done on such topics in the engineering context will also be of value to the medical area. Finally, the profession of medicine is undergoing significant changes in that an increasing number of physicians are becoming employees of organizations of various sizes, and in such capacities are beginning to encounter varieties of ethical problems new to them, but which have been part of the practice of engineering for many years and on which a literature is already developing.

In summary, persons beginning to teach and do research in the field of engineering ethics would do well to explore simultaneously ways in which they could utilize resources from, *and* make contributions to, various related fields such as those discussed above.

XI. Topics for Consideration in Engineering Ethics Courses

There is a broad range of topics that could be touched on in a full-semester course on engineering ethics; indeed there are more than enough for several semesters' worth of discussion. In addition to the obvious topic of the codes of ethics of the various engineering societies, it is usually useful (for reasons discussed elsewhere in this paper) to spend time on a variety of ancillary topics, such as the history and sociology of the profession, organizational dynamics, and interpersonal communications, as well as ethical theory and basic metaethical issues, such as the fact/value distinction. This means that selections must be made from among the numerous possible topics. Unless there are good reasons that dictate otherwise, instructors should be prepared to deal with all possible topics and should leave it to students to select those that are of greatest interest and concern to them.

The following is a partial listing (in no particular order) of some of the more substantial and interesting topics in the area of engineering ethics. Many of the topics listed might not appear at first sight to involve any significant ethical issues, but further consideration should result in an awareness of the more subtle, complex, and significant moral questions associated with each of them. (Some of these topics are elaborated on in more detail elsewhere in this paper.)

 Advertising of engineering services
 Competitive price-bidding

Accreditation of engineering curricula
Engineering licensing and registration
Affirmative action policies in engineering
Age discrimination in engineering
Product liability and quality control responsibilities
Conflicts of interest
Patent rights
Trade secrecy policies
Engineering employment contracts
Problems of engineers employed in foreign countries
Formulation and enforcement of codes of ethics
Setting of fee schedules
Risk assessment
Whistle-blowing
Bribery and kickbacks
Criticism of other engineers' work
Informed consent on engineering projects
Professional autonomy and employed engineers
Technical dissent in organizations
Pro bono activities for engineers
Individual responsibilities on large-scale projects
Unions and professionalism

In line with the assessment of the literature in sections IX and XIII, it must be noted that while there is a minimal literature on most of these topics, there is not an adequate, let alone very strong, literature on any of these topics. In particular, none of them has been discussed in any depth by ethicists, and few of them have been discussed by any nonengineers.

As there are some substantial differences in the ethical issues that arise in the various specialties, it is worth considering setting up separate courses or dividing students into separate discussion sessions in one course to permit them to deal with the topics of greatest personal interest. For example, civil engineers are more likely than electrical, chemical, or others to go into private practice, to join a partnership, to work for a small firm, or to work for a local or state government agency. They are thus more interested in such issues as advertising of services and competitive bidding and less interested in patent rights and dissent in large organizations. Civil engineers are more directly involved and interested in environmental issues than are electrical and certain other kinds of engineers. Civil engineers also have quite different perspectives on such issues as licensing and unionization than do electrical engineers, and this can lead to stimulating (and sometimes heated) debates when both groups are represented in a single class.

XII. Methods for Facilitating the Learning of Engineering Ethics

A number of suggestions for ways of dealing with various topics in the engineering ethics classroom are scattered throughout this monograph, and they will not be repeated here unless necessary. Several other points do need to be made explicit, although space limitations will permit only a brief outline of some of the more important ones.

Persons with experience in any kind of class in any area of "applied" ethics generally agree that the traditional formal lecture format is probably the least appropriate (and least effective) means of dealing with the subject matter. There is less agreement concerning what *is* appropriate (and effective). This is dependent in part on the orientation of the faculty person with regard to the nature of the educational enterprise and also on her/his personality and attitudes toward students. Thus, each instructor will have to work out her/his methods on a trial and error basis. The following suggestions only reflect what several instructors have found to be effective in their own experience.

One of the most frequently used methods for approaching engineering ethics in the classroom is through the use of case studies. This is often useful for the purposes of stimulating discussion, especially if students are asked to participate in role-playing exercises based on specific cases. (The kinds of cases presently available are discussed in section XIII.) However, because most engineering students (with the exception of seniors in co-op programs) have little first-hand (or even second-hand) knowledge of the realities of engineering employment situations, it is often difficult for them to empathize with or even to understand the situations described in the case studies on engineering ethics—especially, early in the course. They can, however, relate quite well to cases, hypothetical or actual, concerning situations encountered by engineering students. Examples include situations in which an engineering student

- seeks help from another student on a take-home assignment
- orders a term paper for an ethics course from a mail-order term-paper "mill"
- believes that an instructor is making unreasonable assignments or is not grading work fairly
- observes another student cheating on an exam.

A valuable exercise—for both the faculty person and the students—is that of negotiating the grading and testing procedures for the course. This not only gives students a sense of participation right from the beginning of the course, but it also strengthens their sense of personal autonomy and responsibility. The discussion of this very concrete practical matter also provides a useful entry into the working out of a common basic vocabulary for the discussion of ethical issues. In the discussion of testing and grading, students inevitably use terms such as "fair," "just," and "right" (in both senses), in addition to talking about "objectivity," "subjectivity," and using other epistemological vocabulary. If the development of each of these concepts is included, the discussion of testing and grading procedures can go on indefinitely and sometimes has to be brought to an arbitrary conclusion. Classes almost always come to a reasonable and workable consensus concerning the requirements for a specific course; if anything, they tend to want to do more work than the instructor can reasonably handle!

Although there are undoubtedly exceptions, for several reasons most engineering ethics courses can incorporate little, if any, explicit consideration of abstract ethical theory. First, it is difficult for even the most competent philosopher to relate (or "apply") ethical theory to concrete cases in engineering ethics, and engineering students are dispositionally opposed, if not openly hostile, to *any* general theory that cannot be directly applied to the solution of concrete problems. (Engineers are *not* physicists!)

Also, there are so many other topics and informational matters that must be discussed that, even if the students had some interest, there would not be much room for a detailed treatment of ethical theory. Just as the presentation of the historical, sociological, organizational, and other aspects of the problems of engineering ethics can and must be made with little or no explicit consideration of historiographical, sociological, and other theories, so the ethical aspects of the problems can and must be done with minimal consideration of the various ethical theories. Of course, interested students can and should be encouraged to take appropriate additional ethics courses.

XIII. Assessment of the Available Literature

A. Textual Material

Although a great deal has been written in the area of engineering ethics, there is a dearth of good material available at present for use in courses dealing with this topic. Almost everything published to date on issues related to engineering ethics has been written by engineers representing the "establishment" position on various issues. Most of these articles, in fact, are written from the perspective of political activists *within* the engineering societies and thus the essays have more of a tone and form of editorial commentaries than they do of scholarly analyses. Only a few engineers, most notably Stephen Unger, Victor Paschkis, Samuel Florman, and Irwin Feerst, have consistently published essays challenging the status quo in the area of engineering ethics. The first essays by philosophers on specific ethical issues in engineering have appeared only in the last year or two, and none of these has yet been published in a readily available outlet. In addition to the lack of breadth of perspective and depth of philosophical analysis in the existing literature, there are also a number of gaps in the literature where significant topics are not discussed at all.[17]

Most of the best articles in the literature have recently been collected into a text anthology, Baum and Flores's *Ethical Problems in Engineering* (1978). There are at present no other books available on the market which are adequate for use in a course on engineering ethics. The only two books on the topic that have been published in the last twenty-five years are both out of print. One of these, Alger, Olmsted and Christiansen's *Ethical Problems in Engineering* (1964), is essentially a collection of brief hypothetical cases and the comments of a number of practicing engineers concerning these cases, generally interpreting them in the light of the various engineering codes of ethics that were extant in 1960. The volume is thus quite dated both in terms of the choice of topics and issues addressed, and also in terms of the analyses given of the cases. The volume is of course also limited in that it is restricted to providing the perspective of a relatively small group of "establishment" engineers in the early 1960s. This volume does provide a useful sense of the state of ethics in the profession twenty years ago, and thus should probably be read by anyone seriously interested in teaching or doing research on contemporary issues in engineering ethics. However, it would not be of any real value as a text for a course in the area of engineering ethics today.

The most important historical perspective on the topic of ethics in engineering is provided by Edwin Layton, Jr.'s, *Revolt of the Engineers*, which it is hoped will soon be available again in paperback. This volume provides a historical analysis of the development of the codes of ethics of the various engineering societies in the larger context of the development of the engineering societies themselves in the early part of the twentieth century. A knowledge and understanding of the basic thesis of this volume and the specific details provided in support of it are almost essential for anyone who would claim to be a competent teacher or serious scholar of engineering ethics. Although highly readable, the book is probably too detailed and sophisticated to be of use with a class of ordinary undergraduate engineering students, but it would certainly be an essential text in any upper-level or graduate course dealing with this topic, even if the basic approach of the course is nonhistorical.

A book which is available, and which might be considered for use as one text in an engineering ethics course, is Samuel Florman's *Existential Pleasures of Engineering*. This is also a highly readable volume, which deals primarily with the nature of the profession of engineering and with the impacts of technology on the larger society. The fundamental ethical points raised in the book, however, and the arguments offered in support of them are relatively brief and straightforward; all are in essence contained in Florman's essay in the October, 1978, issue of *Harper's* which is sufficient for presenting Florman's ideas in the classroom. Nonetheless, this book should be studied in its entirety by anyone proposing to teach a course or to do research in engineering ethics.

B. Case Studies

In addition to the essays and "quasi-editorials" on various topics in engineering ethics, there is only one other general kind of written material—case studies. The case studies in engineering ethics fall generally into two basic categories. One category is a brief, stylized, and relatively abstract example formulated to provide a basis for the discussion of a specific ethical principle, to illustrate an application of a section of one of the codes, or to serve as an example of a generic type of situation commonly encountered by engineers. Some of these brief cases are in fact abstractions from actual cases, while others are fictional or hypothetical. Many of them are quite effective for achieving their specific, but quite limited, purposes. However, such cases are of relatively little interest to most engineering students, and are of little use in discussions of ethics that attempt to go beyond the engineering codes.[18]

The second kind of available case study is more lengthy, detailed, and concretely realistic. These are generally written in a readable journalistic style, and students become interested in them quite quickly. However, they are not as useful in courses as they might at first appear to be, and they have the potential for being used in highly questionable ways by instructors, especially by instructors who have no experience and little understanding of the subtleties and complexities of large technology-based organizations. A survey of the existing literature on whistle-blowing reveals references to a number of "documented" cases of this second kind, citing them as providing evidence of various types of incompetence and corruption at the management level, and also as clear-cut instances of responsible behavior on the part of individual line-engineers. None of these case analyses show any recognition of the fact that most of them are presented (and "documented") only from the single perspective of the line-engineer. It is very difficult to deal with these one-sided accounts in the classroom, but at the very least the one-sidedness of the presentation must be pointed out as explicitly as possible.

While it is doubtful that any complete and balanced case studies will be available for use in courses, some cases must be recongized as being better than others, and certain procedures can be followed in class discussions to compensate for some of the inadequacies of the cases. Several cases are reasonably well documented in government reports and transcripts of congressional hearings. For example, the B.F. Goodrich A7-D Brake Case,[19] which has been publicized solely from the perspective of one party to the case, is presented from a variety of other perspectives, including the Air Force, GAO investigators, and the company in testimony before a joint House-Senate investigating committee. Appropriately selected excerpts from the hearing transcript can provide students with a reasonably balanced account of the case, and instructors can prepare for class discussion by studying the complete transcript. Similarly, a team of researchers at Purdue University has compiled a comprehensive set of materials concerning the case of the three engineers fired by the Bay Area Rapid Transit (BART) system management. This set provides instructors with a wealth of relevant information to supplement the briefer and less complete and balanced reports that are presently available for use in engineering ethics courses.[20]

In using any of these case studies, instructors should try to devote a substantial part of the discussion to identifying information, missing from the accounts, that would be relevant to any moral judgments about the case. Every effort should be made to make students aware of the potential biases in any single account of a case. If possible, the cases should be used primarily as the basis for class discussions of what students would (and morally ought to) do should they find themselves in similar situations, rather than trying to make judgments about the cases themselves. It is also helpful to spend some time discussing what might be done, either by individuals or by the engineering profession or even society at large, to prevent similar situations from recurring in the future. Finally, some effort should be made to assess the actual frequency of occurrence of such situations, with attention being paid to the ways in which a specific case may be an exception rather than the rule.

XIV. Research Needs and Priorities

The lack of a literature and thus of adequate materials for use in engineering ethics courses is in and of itself a sufficient justification for encouraging new research in this area. There are other justifications, but the *need* for more and better research on the whole range of issues in engineering ethics will not be belabored here.

Although thoughtful analysis of almost any topic related to engineering ethics would be a welcome addition to the literature, there are three basic kinds of work which would be of particular value, and to which the highest priority should be assigned, in order to make the greatest amount of the high-quality material available as soon as possible for both students and teachers. The three general categories are: (1) topics which have been dealt with in the existing literature of engineering ethics, but which have been given grossly inadequate treatment; (2) topics which have not been dealt with to date in the engineering ethics literature, but which have been dealt with in significant ways in the literature of other professions; and (3) topics which have been essentially ignored in the literature, not only of engineering ethics, but also of all of the other professions. I shall look briefly at only one or two examples of topics from each of these categories, outlining some of the directions in which the new literature might be developed.

A. Whistle-blowing

A prime example of a topic which has been treated, but very inadequately, in the engineering ethics literature is the topic of whistle-blowing. Almost everything published on the topic to date focuses on the need for the development of new mechanisms and procedures for protecting engineers whose careers have been placed in jeopardy for having gone outside the normal channels in order to do what the individual engineer perceived as protecting the public from potential risks. Some articles have also been written that discuss alternative mechanisms that could be established in large organizations to help resolve conflicts satisfactorily, thus eliminating the need for engineers to "blow the whistle" outside the organization.[21] (It should be noted that the general literature on whistle-blowing does not offer much more than the articles restricted to engineering.)

The discussions of these aspects of whistle-blowing are certainly significant, but there are a number of other aspects of the topic which merit consideration, not only from a purely scholarly perspective, but also as being essential for dealing adequately with this complex and important topic in the classroom. An adequate treatment of these additional aspects requires contributions from ethicists as well as from social and behavioral scientists and management experts. Four of the more important and interesting aspects of the general topic of whistle-blowing will be outlined here to suggest some directions for future research.

One of the topics most relevant to the problem of whistle-blowing is that of collective responsibility. Whistle-blowing, by definition, can occur only in an organizational structure, insofar as it involves going outside the normal organizational channels to report to persons not part of the organization. Thus, one of the central questions that must be answered in dealing with both general and specific whistle-blowing cases is that of when the individual has a responsibility, either to the organization or to some other group, which justifies overriding the judgment of the organization itself. Many of the extant discussions of whistle-blowing have argued that in the long run whistle-blowing is, in fact, in the best interest of the organization, but not much evidence, empirical or theoretical, is given in such discussions. Most important, none of these discussions even consider the aspects of the problem that have been discussed in the philosophical literature over the last decade with regard to assigning moral responsibilities to groups or organizations collectively, and the problems of assigning responsibilities to individuals who are members of large organizations. I would like to suggest that such an effort to relate the existing philosophical literature on collective and group responsibility to the problem of whistle-blowing in engineering organizations would not only make a significant contribution to the field of engineering ethics, but would also result in new insights that would lead to further refinements and even perhaps to major reformulations of the current philosophical theories on collective responsibility.

The second issue that would benefit greatly from consideration of the existing philosophical literature is that of the problem of conflicting beliefs or truth-claims in engineering contexts. The essence of almost every documented engineering whistle-blowing case is that of disagreement between two individuals concerning a technical judgment. There is no awareness whatsoever shown in any of the discussions of these cases or of the general topic that an extensive philosophical literature exists both on the general epistemological problem of criteria for scientific truth or validity and the somewhat more specific topic of the establishment of probabilities. Most discussions tacitly assume a relatively naive form of positivism or logical empiricism. The assumptions—not only tacit, but sometimes even explicit—concerning probability theory are often astonishingly naive, as though probabilities are in some way directly read out of sense experience. Thus, the few existing discussions of whistle-blowing that relate to the question of when an individual has a sufficient degree of certainty, or when the individual's belief is adequately justified to make it reasonable for that person to press his or her belief to the point of ultimately blowing the whistle, is at present quite inadequate and in need of substantial philosophical "beefing-up." At the very least, a consideration of the post-Kuhnian work in philosophy of science and the work in the area of Bayesian subjectivist theories of probability would greatly enrich the whole discussion of the issue of whistle-blowing.

The consideration of the general problem of whistle-blowing from the broader philosophical perspective of the discussions of group responsibility and epistemology leads to the third point, which has not been discussed at all either in the context of engineering ethics or in the general philosophical literature. An examination of the documented cases of whistle-blowing supports the belief that the person who ultimately blew the whistle in any given case was essentially locked into a sequence of events from a very early stage in which the whistle-blowing was an almost inevitable outcome. The parties in the cases—both the whistle-blowers and the involved management persons—held naive and often conflicting views of the nature of the relation between individual and group responsibilities and the problems of justified belief and knowledge. In each case it appears that the whistle-blower believed from the very beginning that he or she knew all of the relevant facts with a high, if not absolute, degree of certainty. The situations were usually the worst in cases where management held a contrary belief with a similar assumption of truth and certainty, although even in the cases where management was willing to recognize its own potential fallibility, the insistence of the staff engineer on the certainty of his or her position made communication and reasonable resolution of the conflict almost impossible. Recognition of the role of epistemological assumptions in whistle-blowing cases brings out the relative futility of establishing any kind of a *mechanism* for satisfactorily resolving conflicts at a stage before the individual feels it is necessary to blow the whistle. As long as the individual believes that he or she is absolutely right, it is irrelevant how many alternative channels and levels of appeal may be available to that person.

In most documented cases of whistle-blowing, the individual engineer has reported what he or she perceived to be a serious threat to the health and safety of others through at least several layers of the organizational hierarchy. In few cases did the engineer seek or get independent competent outside evaluation of his or her beliefs, either before or during the process of pressing the matter with management. In such cases, the engineers assumed that the failure of management to agree with their judgment indicated either that the persons in the managerial positions were incompetent to make an intelligent judgment on such a matter, or that the managers were "corrupt" in the sense that they were solely concerned with maximizing short-term financial gains. Such an attitude put management in a defensive posture and almost inevitably led to the shifting of the concern with the crucial issue, the potential hazards to health and welfare, to defending the integrity of individual managers against what were perceived as personal attacks. These problems cannot be adequately analyzed in this context. It is sufficient to remark that much remains to be said concerning the potential whistle-blower's responsibility to present her or his concerns to management in such a way as not to impugn the professional or personal reputation of managers and to do everything possible to keep the attention focused on the substantive issues, rather than allowing discussions to be sidetracked into consideration of personalities. An equal amount of effort needs to be devoted to the management perspective to better equip managers to deal with such situations, when and if they arise. (This is an area for fruitful collaboration between business and engineering ethics.)

Once again, as long as the individual believes that he or she is right, and that anyone that disagrees must be either incompetent or corrupt, it does not matter how many alternative channels of appeal may be made available *within* an organization. The one strategy that might work for dealing with such cases would be to present the substantive issues to an outside consultant, acceptable to both the individual engineer and management, to adjudicate the disagreement. The feasibility and effectiveness of such a strategy is something which again requires considerable study. It is a service which possibly could be provided by the appropriate engineering professional society.

The tendency of whistle-blowers to perceive management persons as being incompetent to evaluate technical issues and/or as being concerned solely with short-range profits reveals a fourth aspect of the whistle-blowing problem that also deserves further study and consideration. It is an undeniable fact that the individual engineer who believes that he or she has identified a problem that could result in harm to others is in a somewhat more vulnerable position than his or her supervisor. It is a fact, recognized by most staff engineers, that supervisors are seldom, if ever, held accountable for the negative impacts of decisions that they made more than a year or two earlier. In most organizations, the supervisors move either upward or horizontally within the organization or move to other organizations on a fairly frequent and regular basis. Thus, if an engineer makes a recommendation that a certain product be redesigned because of potential harm to users, and this recommendation is overridden by the supervisor, if the effects of that decision are not identified for at least several years, the supervisor either will have been promoted in the organization and thus be in a position to protect himself or herself, or will have moved into some other position remote enough that he or she will feel little, if any, negative consequences. This fact encourages the staff engineer to believe or to assume from the beginning that no one in management is seriously concerned about the public welfare, and thus encourages the engineer to take something of an adversarial position with his or her supervisors at a very early stage.

Thus, another potentially significant area of investigation would be that of the impact or consequences of attaching criminal liability to managers who override or countermand the recommendation of staff engineers, in cases where the staff engineer has reported potential harm that might result from following that strategy. It seems quite reasonable that attaching such criminal liabilities would make management more cautious in overriding staff engineers' recommendations and would also increase the engineers' willingness to trust the ultimate judgment of their supervisors. This might result in a more cooperative atmosphere within engineering organizations. However, the issue can also be considered from a more abstract and less consequential perspective as simply assigning responsibilities in a reasonable and just manner. Certainly, if a manager makes a judgment that eventually results in harm being done to other individuals, and this harm had been predicted by a technically competent engineer, it would be difficult not to see such an act as involving at least negligent, if not intentional, battery. The reasonableness of such a policy stands in need of further careful consideration from the perspective of ethics and social and political philosophy.

It should be noted that none of the changes in policies and procedures for "protecting" whistle-blowers will have much practical impact as long as the underlying legal system maintains the primacy of employers' rights to terminate employees at will, without cause. Most remedies presently under discussion only protect employees from being fired for having explicitly blown the whistle. The employer can discharge the employee as long as it is not explicitly or blatantly in response to an act of whistle-blowing. Thus, the basic issue of the right to job security, as opposed to the employer's almost unlimited right to fire, needs to be carefully reexamined from both ethical and jurisprudential perspectives.

One final area where work might be done with regard to the problem of whistle-blowing is in terms of an analysis and reevaluation of the codes of ethics of the various engineering societies. Other than rather general platitudes about "holding paramount the general welfare" or informing the public of harmful products, the codes are still excessively vague on this issue. It is to be hoped that the new procedures, which are being established by the various societies, will in coming years result in refinements to, and elaborations of, the codes in ways which will be useful also for the teaching of engineering ethics. One such "model" code that more explicitly supports engineers' rights and responsibilities to communicate significant information is presently being developed by the Committee on Scientific Freedom and Responsibility of the American Association for the Advancement of Science.

B. Informed Consent

Probably the most significant topic which falls into the category of issues dealt with in significant ways in the literature of other professions but which have not been discussed at all in the context of engineering ethics, is the topic of informed consent. The literature in engineering ethics has consistently followed the highly paternalistic approach embedded in the codes of the various engineering societies. The basic codes of almost all the engineering societies include assertions to the effect that "the engineer should hold paramount the public health and safety," and almost all of the essays written by engineers on ethical issues are concerned with the question of how engineers should hold public health and safety as their primary concern. The essays either deal with an answer to the question, "Is such and such, in fact, in the best interest of the general public?" so that on answering this question the engineer will know what he or she should or should not do in professional practice, or with specific cases in which an engineer *knows* what is in the public interest and is concerned then with determining the best means for acting to promote the public welfare.

Only a few engineers, most notably Samuel Florman, have challenged the paternalism and "expertism" implicit in these principles and case analyses, but this questioning has led only to the conclusion that since the engineer has no special qualifications for determining what is in the general welfare, he or she is responsible only for what he or she is assigned to do by an employer or client. No engineers have seriously considered whether or not they have a responsibility similar to that of physicians, scientific researchers, and other professionals to obtain the informed consent of parties to be affected by their actions before embarking on those actions, although it appears prima facie that there is no reason why engineers should not bear this as a responsibility associated with their professional practice. There are indeed some significant differences between the physician or scientific researcher and the engineer, but the fundamental moral obligation would seem to be the same in all cases. Unlike the physician or researcher, the engineer does not usually deal personally with those who will be affected by her or his professional activity and is almost never involved in an activity where individuals are affected singly. The highway or airplane designed by the engineer will affect the health and welfare of numerous individuals, few of whom the engineer will ever meet. Significant problems thus arise both as to how the engineer can inform persons ultimately to be affected by her or his work of the potential consequences of that work, and even more serious problems are associated with working out ways of obtaining consent from those persons. The literature in the areas of medical treatment and experimentation is certainly a very useful starting point for consideration of the questions of communication, information, and acquiring consent, and it is therefore important that persons who have worked in these areas and are familiar with this literature become involved in the examination of these questions in the new context of the profession of engineering.[22]

C. Discrimination and Affirmative Action

One substantive topic in the area of engineering ethics which has been dealt with in related fields, such as law and medicine, but which involves radically different problems in the context of engineering, is the issue of discrimination and affirmative action policies. As indicated in the general description given earlier of the nature of the profession, engineering is today one of the most male-dominated of any of the employment categories in the United States. It is important to recognize, however, that because of changes in attitudes on the part of the engineering establishment over the last several years (whether as the result of social or governmental pressures, or simply from the recognition of the basic unjustifiability of the previous attitudes and policies in the profession), there are now many more openings for women in engineering schools, and especially in employment situations, than there are women interested in filling those positions. The problem in engineering is not related in any significant way to the issues raised in the Bakke case or any of the other more publicized cases concerning affirmative action programs. Instead, the ethical issues associated with the affirmative action programs in engineering are issues that have not even been publicly identified, let alone carefully discussed and analyzed. Only a brief sketch of the issues can therefore be given here, primarily for the purpose of indicating the ways in which these issues at present are relatively specific to the field of engineering.

The affirmative action programs now being pursued to increase the number of women in the engineering profession are following either or both of two basic strategies—each of which raise serious ethical questions. It is widely accepted that the profession of engineering is not attractive to women, generally because of its public image as involving intensive, narrowly focused, highly quantitative technical activities. Operating on this assumption, efforts are being made to change the public image of the nature of the profession and also to change the self-image or attitude of young women, both strategies being designed to match more closely the image of the profession with the interests and attitudes of women who would consider it as a possible career. Unfortunately, the public image of the nature of the engineering profession is relatively accurate, especially with regard to describing the nature of the undergraduate educational program required for entry to the profession and for the first years of professional activity following graduation. As indicated earlier, the undergraduate curriculum in engineering *is* highly restrictive and highly demanding, and the persons who "survive" the program are generally individuals whose personality, attitudes, and interests are also highly limited and restrictive. It is morally questionable to change the image and to provide a different picture of the nature of the educational and professional activities of an engineer, that is, to misrepresent the facts of the situation, for the purpose of trying to attract more women into the profession. Likewise, it is as questionable, if not more so, to try to change the personalities or attitudes of women or members of minority groups. The only remaining option is that of changing the curriculum to fit the personalities and attitudes of a more diverse group of individuals in such a way as to increase the diversity in terms of sex and ethnic groupings.

In brief, a critical consideration of the issue of affirmative action policies in engineering from an ethical point of view would ultimately involve the consideration of a whole range of significant, related issues, including behavior modification and accreditation policies and procedures, as well as the basic question of the goals and justifications for increasing the number of women and minorities in the entire range of professions.

D. Advertising

The third area which should be given priority for the development of new classroom materials (and essential for establishing the base for new research) is that of the topics that have not been dealt with in any significant way to date in the literature devoted to any of the fields of applied ethics. One example of such a topic is that of the advertising of professional services. Although this issue was widely discussed in both engineering literature and the literature of many of the other professions before the 1976 Supreme Court ruling forbidding the proscription of advertising by professional societies, the discussion was limited entirely to the question of whether advertising should be permitted at all. Today the primary issues are related to the question of what kind of advertising is ethically justifiable and what kind is unethical. There is a substantial literature that has developed over the years in the area of business ethics, but this literature contains no significant input from philosophers and others outside the field of advertising itself. It is also restricted to the issues associated with product advertising, many of which are similar to, but many of which are different from, the issues associated with the advertising of professional services.

The critical issues in the advertising of professional services are tied to the uniqueness of each individual case with which the professional (engineer, lawyer, physician, or journalist) deals. The basic principle of product advertising—that the advertisement be factually verifiable—is not adequate for dealing with professional services. The lawyer who advertises the true fact that he or she has won 94 percent of criminal cases taken before a jury may be highly selective in the choice of cases that he or she handles, and therefore may be no more competent or effective than a lawyer who had only a 50 percent success rate, but who accepts any and every case that comes through the door. The problems here are not restricted to a single profession, but cross all the professions. Thus, once again, the work that needs to be done should not be carried out exclusively by those interested in the area of engineering ethics, but should be done by representatives of many different professional fields collaborating and interacting with philosophers, historians, and others who are interested in the general issues common to all of the professions, as well as to specific professions.

Other topics, common to the various professions (and also often to business), include competitive price bidding for professional services, control of accreditation of professional programs, control of licensing and registration activities, and the setting of technical standards in specific areas. Another issue which has not been discussed at all in any of the literature to date, and which can only be discussed in a multiprofessional context, is that of the assignment of responsibilities to various professionals working in collaborative enterprises. For example, it is not unusual at present to find a number of different professionals—including physicians with a variety of different specializations, engineers, nurses, technicians, therapists, psychologists, and administrators—all working together on a single project, sometimes involved in the treatment of a single patient or client. Questions of collective responsibility and of the assignment of specific responsibilities to individuals involved in such collaborative efforts have not been discussed in any significant way in any of the literature. While the discussion of such issues is important in the context of ethics, even more impetus for the further study of such questions is being generated by the increasing number of legal actions arising out of such collaborative activities.

XV. Conclusion

There is no question that the issues in engineering ethics are intellectually challenging and socially significant, and that scholarly activities dealing with these issues should be encouraged. The only difficult question is that of how to attract sincere and competent individuals from a variety of disciplines to work together in a constructive way, as required by the complexity of the issues. Fortunately, such persons are beginning to come together more or less spontaneously, and a small community of scholars is now coming into existence. If they are at all successful in dealing with even a few of the issues identified in this paper, the long-term success of the field and the place of engineering ethics in the curriculum should be assured.

Notes

1. Joseph M. Dasbach, *EVIST Resource Directory* (Washington, D.C.: AAAS, 1978). This study confirms the trend towards the establishment of this field reported in Ezra D. Heitowit, *Science, Technology and Society: A Guide to the Field* (Program on Science, Technology and Society, Cornell University, 1976) and Ernest M. Hawk, *Technology and Society Courses at the College Level* (INPUT Program, Pennsylvania State University, 1976).

2. A very good *comparative* analysis of the *current* engineering society codes is made by Andrew G. Oldenquist and Edward E. Slowter, "Proposed: A Single Code of Ethics for All Engineers," *Professional Engineer,* May 1979, pp. 8–11.

3. Samuel Florman, *The Existential Pleasures of Engineering* (New York: St. Martin's Press, 1976).

4. The most recent is "Moral Blueprints: On Regulating the Ethics of Engineering," *Harper's,* October 1978.

5. The IEEE Member Conduct Committee's report on this case was published in *Technology and Society,* no. 22 (a newsletter of the IEEE Committee on Social Implications of Technology), Spring 1979.

6. "Engineering College Instruction in Professionalism—A Survey of Faculty Attitudes" (Washington, D.C.: NSPE, 1963).

7. "Instruction in Professionalism—A Survey of Attitudes of Engineering Faculty" (Washington, D.C.: NSPE Publication no. 2008, 1971).

8. John A. Bonell, ed., "A Guide for Developing Courses in Engineering Professionalism" (Washington, D.C.: NSPE Publication no. 2010, 1976).

9. The most ambitious of the projects designed to develop the field of engineering ethics and to stimulate interaction between engineers and philosophers is described in Albert Flores, "National Project on Engineering Ethics to Bring Together Engineers, Philosophers," *Professional Engineer,* August 1977, pp. 27–29.

10. One of the most fully developed of these new courses is described in Vivian M. Weil, "Moral Issues in Engineering: An Engineering School Instructional Approach," *Professional Engineer*, October 1977, pp. 45–47.

11. Bonell, "A Guide for Developing Courses...," p. 5.

12. Billy T. Sumner, "Get 'Em Young and Bring 'Em up Right!" *Consulting Engineer*, November 1975, pp. 20–22.

13. Some of the specific legal actions are outlined in Milton F. Lunch, "Public Accountability for the Professions," *Professional Engineer*, October 1978, pp. 16–18.

14. These goals and arguments on their behalf are spelled out in substantial detail in *The Teaching of Ethics in Higher Education: A Report by The Hastings Center* (Hastings-on-Hudson, N.Y.: The Hastings Center, 1980).

15. A description of one project is provided in Deborah G. Johnson's "Computer Ethics: New Study Area for Engineering, Science Students," *Professional Engineer*, August 1978, pp. 32–34.

16. This judgment is confirmed by Robert Veatch's article, "Codes of Medical Ethics: Ethical Analysis," in the *Encyclopedia of Bioethics* (New York: The Free Press, 1978).

17. A comprehensive bibliography on engineering ethics, prepared by Robert Ladenson of the Illinois Institute of Technology, is scheduled for publication in late 1979. A partial selective bibliography on this topic was published in *Business and Professional Ethics* 1, no. 2 (Fall, 1977).

18. The most substantial collections of these cases are the compilations of the *Opinions of the Board of Ethical Review*, published periodically by the National Society of Professional Engineers. Vol. 4 was published in 1976.

19. One of the principals presented his personal account in Kermit Vandivier, "The Aircraft Brake Scandal," *Harper's*, April 1972, pp. 45–52. The congressional hearing was held before the Subcommittee on Economy in Government of the Joint Economic Committee, December 13, 1976.

20. Anderson, Robert M. et al., eds. 5 BART Case Studies (Washington, D.C.: American Society for Engineering Education, 1979).

21. The most comprehensive studies on whistle-blowing are those of Rosemary Chalk and Frank von Hippel, "Due Process for Dissenting Whistleblowers," *Technology Review* (June/July 1979), and Sissela Bok, "Whistleblowing and Professional Responsibilities," in Daniel Callahan and Sissela Bok, eds., *Ethics Teaching in Higher Education* (New York: Plenum Press, 1980).

22. The two sides of this issue are debated in Robert Ladenson's paper, "The Social Responsibilities of Engineers and Scientists: A Philosophical Approach," and Robert Baum's paper, "The Limits of Professional Responsibility," both of which are included in *Values and the Public Works Professional*, to be published by the American Public Works Association.

Bibliography

Alger, Philip L., Christensen, N.A., and Olmsted, Sterling P. *Ethical Problems in Engineering*. New York: John Wiley, 1965.

Anderson, Robert M. et al., eds. *5 BART Case Studies*. Washington, D.C.: American Society for Engineering Education, 1979.

Baum, Robert J. and Flores, Albert W. eds. *Ethical Problems in Engineering*. Troy, N.Y.: Rensselaer Polytechnic Institute, Center for the Study of the Human Dimensions of Science and Technology, 1978.

Beauchamp, Tom L., and Bowie, Norman E., eds. *Ethical Theory and Business*. Englewood Cliffs, N.J.: Prentice-Hall, 1979.

Bonnell, John A., ed. *A Guide for Developing Courses in Engineering*. NSPE Publication 2010, Washington, D.C.: National Society of Professional Engineers, 1976.

Business & Professional Ethics: A Quarterly Newsletter/Report. Troy, N.Y.: Rensselaer Polytechnic Institute, Center for the Study of the Human Dimensions of Science and Technology, 1977–.

Callahan, Daniel and Bok, Sissela, eds. *Ethics Teaching in Higher Education*. New York: Plenum Press, 1980.

Chalk, Rosemary, and Von Hippel, Frank. "Due Process for Dissenting Whistle-blowers," *Technology Review* 81, no. 7 (June/July 1979).

Collins, Randall. *The Credential Society*. New York: Academic Press, 1979.

Conference on Engineering Ethics: Proceedings of a Conference held in Baltimore, May 18–19, 1975. New York: American Society of Civil Engineers, 1976.

Dasbach, Joseph M. *EVIST Resource Directory*. Washington, D.C.: American Association for the Advancement of Science, 1978.

Ethics, Professionalism & Maintaining Competence: Proceedings of a Con-

ference held at The Ohio State University, March 10–11, 1977. New York: American Society of Civil Engineers, 1977.

Ewing, David W. *Freedom Inside the Organization.* New York: Dutton, 1977.

Florman, Samuel. *The Existential Pleasures of Engineering.* New York: St. Martin's Press, 1976.

———. "Moral Blueprints; On Regulating the Ethics of Engineers," *Harper's,* October 1978.

Fruchtbaum, Harold, ed. *The Social Responsibility of Engineers.* The Annals of the New York Academy of Sciences, vol. 10, art. 10 (Proceedings of a symposium held in 1972).

Hankins, George. "Evaluating Student and Faculty Attitudes Toward a Course in Technology and Society." *Engineering Education,* February 1977, pp. 400–2.

Hawk, Ernest M., ed. *Technology and Society Courses at the College Level.* Report from the Increasing Public Understanding of Technology Program, Pennsylvania State University, 1976.

Heitowit, Ezra D. et al. *Science, Technology and Society: A Guide to the Field.* Program on Science, Technology and Society, Cornell University, 1976.

Heitowit, Ezra D. *Science, Technology and Society: A Survey and Analysis of Academic Activities in the U.S.* Program on Science, Technology and Society. Cornell University, 1977.

Knepler, Henry. "The New Engineers." *Change,* June 1977, pp. 30–35.

Larson, Magali S. *The Rise of Professionalism.* Berkeley: University of California Press, 1977.

Layton, Edwin T., Jr. *Revolt of the Engineers: Social Responsibility and the American Engineering Profession.* Cleveland: Case Western Reserve Press, 1971.

Lynn, Walter R. "Engineering and Society Programs in Engineering Education," *Science,* 14 January 1977, pp. 150–55.

Mathes, J.C., and Chen, Kan. "Educational Objectives for Science, Technology and Society Programs," *IEEE Transactions on Education* E–21, no. 1, February 1978, pp. 27–30.

Nader, Ralph, Petkas, Peter, and Blackwell, Kate, eds. *Whistle Blowing.* New York: Grossman, 1972.

Opinions of the Board of Ethical Review 4. Washington D.C.: National Society of Professional Engineers, 1976.

Perrucci, Robert and Gerstl, Joel E. *Professional Without Community: Engineers in American Society.* New York: Random House, 1969.

Roy, Rustum, ed. *A Survey of Academic Technology and Society Activities.* Report from the Increasing Public Understanding of Technology Program, Pennsylvania State University, 1976.

Sumner, Billy T. "Get 'Em Young and Bring 'Em up Right!" *Consulting Engineer* 45 no. 5, November 1975, pp. 20–22.

The Teaching of Ethics in Higher Education: A Report by The Hastings Center (Hastings-on-Hudson, N.Y.: The Hastings Center, 1980).

United States Congress, Subcommittee on Economy in Government of the Joint Economic Committee, *Hearing on the Air Force A–7D Brake Problem* (Dec. 13, 1976). Washington, D.C.: U.S. Government Printing Office, 1976.

Vandivier, Kermit. "The Aircraft Brake Scandal," *Harper's,* April 1972, pp. 45–52.

Weil, Vivian M. "Moral Issues in Engineering: An Engineering School Instructional Approach," *Professional Engineer,* October 1977, pp. 45–47.

Weinstein, Deena. *Bureaucratic Opposition: Challenging Abuses at the Workplace.* New York: Pergamon Press, 1979.